The Collecting Bug

A Rock Collector's Philosophy of Life

Bob Farrar

FOREWORD

There are many people in this world who get the urge to collect things, an urge that might be called the "collecting bug." This book is about my particular collecting bug, the hobby of collecting of rocks, minerals, fossils, gemstones, and the like. However, much of what I have included could be applied to the collection of other things, both manmade and natural; rocks are simply the object of much of my own efforts. There have been many books published on this subject. There are numerous excellent works on what to collect, where to collect, and how to identify what you collect. What I intend to explore is, rather, *why* - and why not – to collect. It is my hope that the reader will gain an appreciation of avocational rock collecting and why so many of us pursue it.

Where appropriate, I will refer readers to other sources of information. In the text, these references are indicated by the author's last name, or the name or initials of organizations, and the date published. The full citations are then listed in the bibliography.

CONTENTS

ACKNOWLEDGMENTS

I would like to thank my parents, Elizabeth and Russell Farrar, for their patience and encouragement of my rock collecting hobby, as well as my fourth-grade teacher, Georgia Erving, for getting me started collecting, and the many other teachers that influenced my life. Thanks also to the many fellow collectors that have shared their knowledge with me over the years. Thanks to the many people who have helped me to visit productive collecting sites, with a special thanks to Mary Groves of Maine. And, thanks to Sara Mount, through whose efforts I have been able to travel to many famous gem, mineral, and fossil localities. Thanks to Pixabay.com for pictures used in the cover design. Finally, I would like to thank my late wife, Lisa, for her unending love and support.

Chapter 1

Introduction

I have been collecting rocks, minerals, and fossils for most of my life, which now spans several decades. This activity has been mainly a hobby for me, not a part of my professional career, though I hope to convey how the two are related. There have been many fine books published with the amateur rock collector in mind. There are numerous excellent references that give descriptions of the characteristic used to identify minerals and fossils. Likewise, there are many books that list sites where various rocks can be collected. Many of these works also give helpful information on how to extract minerals or fossils from the ground, and how to properly clean, prepare, catalog, and display them. A quick search of the website of any major bookseller will turn up numerous such books. It is not my intention to rehash material that has already been thoroughly explored by other authors. It is, rather, to explore the question of *why* to engage in the hobby of collecting rocks, and why to expend so much time, effort, and money as I and many others have done.

The focus of this work is strictly on the *hobby* of rock collecting. It is not my intention to advise those considering geology or paleontology as a profession. In some cases, though, those who begin as hobbyists may wind up choosing to become professionals.

Humans have been collecting things as long as we have been on this Earth. For most of our history, collecting things, particularly edible things, or things that could be used as tools or shelter, was essential to our survival. Humans may thus have an innate instinct to collect things. As acquiring food and other essentials has become easier, this instinct can be manifested in the desire to collect other things (see, for example, Catchpole 2010). Herein, I focus on my particular desire to collect rocks, but the same principles can be applied to many other collecting activities.

I have a deeply rooted passion for knowledge. In my case, I am particularly interested in the natural world, while others with a similar passion may be more interested in the works of mankind. This passion extends from generating new knowledge to the preservation and dissemination of both new and existing knowledge. This is why I chose to make my living as a scientist, rather than choosing a career in which I might make lots of money. My professional life has had nothing directly to do with rocks and minerals. It has, though, lead to some modest discoveries and the publication of a number of papers that have been circulated widely in their field.

Over the years, I have had the opportunity to visit some of the many mineral and fossil localities of Brazil and Morocco. I have found that there is considerable misinformation, misunderstanding, and ignorance about these localities, even among knowledgeable rock collectors. To help correct this, I have endeavored to disseminate what I believe to be accurate knowledge that I have gained. I have done this primarily through a number of articles that I have published in *Rock & Gem* magazine, as well as talks given to local clubs and articles in their newsletters. While I will touch on some of this work herein, I will refer readers to those articles where appropriate, rather than repeat too much.

A major point that I will make, though, is that, to me, rocks, minerals, and fossils themselves constitute knowledge. When properly documented, a specimen of, say, quartz, provides a record of the fact that quartz occurred at a particular site. In sufficient quantity, information like this can aid in the understanding of the geological forces that shape the Earth, and the development of economically important mineral deposits. Such information is also particularly important for fossils, as it can aid in the construction of evolutionary histories. There are those who will argue that collection of this type of knowledge is best left to professionals. I will be touching on this argument later, but suffice it to say here that there are simply not enough professionals to collect every bit of potentially useful knowledge.

I know that there are others who share the same idea that mineral specimens constitute knowledge, though I have rarely heard it in so many words. A similar philosophy is evident in the title of a special exhibit at the Munich, Germany, mineral show in 2019: *Wer sammelt, schreibt Geschichte.* This roughly translates to "Whoever Collects, Writes History," with which I totally agree.

As I will go into in more detail, there are more reasons than generating knowledge to collect rocks. It can be a very educational activity, for both adults and kids. One can learn a lot about the world by studying rocks. Many rocks and minerals are simply beautiful, and can be a pleasure to look upon. An appreciation of beauty of rocks can lead one into gemstones, gem cutting, and jewelry making. One can even earn a little money. But, as I will discuss later, making money is not, in itself, a particularly good reason to become a rock collector.

One aspect of rock collecting that I will not be discussing is the metaphysical aspect. There are already numerous publications about crystal healing and metaphysics.

One thing that you learn in science, no matter what field, is that you cannot publish something on which you or someone else has published already. You also learn to be as concise as possible. That is one reason I am not going to go into detail on how to find or identify rocks, or how to cut gemstones or make jewelry. There are already many good books on these topics. Others have also published on reasons to collect. However, unlike identifying minerals, this is a subjective area, on which everyone's point of view differs. Subjectivity is one way in which this will differ from scientific writing, where we are taught to be as objective as possible. It is my hope to convey my own synthesis, in hope that the reader may find it useful in their pursuit of the hobby. In addition, while my discussion will focus on rocks, minerals, and fossils, the same principles can be applied to just about anything that can be legally and ethically collected, both natural and manmade.

Chapter 2

Beginnings

There are many ways by which people come into the hobby of rock collecting. In my case, I credit (or blame!) one of my fourth-grade teachers. Knowledgeable teachers with a passion for passing on their knowledge to their students are responsible for influencing the vocational and avocational choices of countless students in all fields. This is true for me in my professional life. In many cases, a teacher's influence may also spark a lifelong interest in a subject not directly related to one's professional life. Such was the case with me. I remember many of my fourth-grade classmates also showing interest in rocks, but I think that I may be the only one to have stuck with it ever since.

Next to my teacher, I must also credit my parents. Neither of them had much knowledge of rocks and minerals, but they encouraged my pursuit of the subject. They indulged my interest endlessly, taking me to local collecting sites and shows, buying me small specimens, and eventually, letting me have some very basic rock polishing equipment. And I'm not the only one. Over the years, I have met many enthusiastic collectors who inherited their interest from their parents. And it is not always the parents who influence their kids; sometimes it is the other way around. Just look at how many kids know more about dinosaurs than their parents do.

This brings up a major reason why people pursue rock collecting – it is something that families can do together. There have been many times that I have seen parents with their kids out collecting, perusing a rock show, or visiting a museum. Even if it does not lead to a long-term interest in rocks, these experiences help instill an interest in learning and an appreciation for the natural world. Oftentimes, the parents agree to doing this to indulge their kids, only to wind up enjoying the experience as much or more themselves. I've seen parents digging for stones, getting all muddy, and having as much fun as the kids. However, it always seems that at the end of the day it is a kid who finds the best specimen.

There are many other ways that people come into this hobby. One may have an enthusiastic friend who is into it. A person may go to a gem and mineral show looking for jewelry to give as Christmas presents and wind up looking at the minerals as well. Those with an artistic bent may follow an interest in jewelry design into gemstones and gem cutting. Those with an interest in biology may follow that interest into an interest in fossils. A visit to a good museum can spark the interest. And, a simple appreciation for beautiful things can lead one into rock collecting. The list goes on.

Once a person has developed an interest in rocks, minerals, fossils, or the like, what does one need to do to get started? Read, that's what. I would suggest getting a good basic book on the subject. Many people start with *A Field Guide to Rocks and Minerals* (Pough 1997). However, there are many other choices. Any book with information about physical characteristics of rocks and minerals, how to distinguish them, how they are formed, and where they are found can be helpful. Today, there is also a vast amount of information available online. One word about internet searches, though; often, when one searches the name of a mineral, the first several hits may be for websites of people selling that mineral. Sites with useful information about it may be down the list.

I personally find books that are mostly pictures to be of limited value. While pictures can be beautiful to look at, they may not be very helpful in identifying rocks and minerals. For other subject areas, such as birds or butterflies, pictures are extremely helpful. This is because there is not so much variation between specimens. Every adult bald eagle, for example, looks a lot like every other adult bald eagle. Some species of birds have differences between sexes or adults and juveniles, but these are limited. Every specimen of a mineral, by contrast, is slightly to very different form every other specimen. Calcite, for example, occurs in hundreds of crystal forms, not to mention a huge range of colors and sizes. Most general books on rocks and minerals have only one or a few pictures of each mineral. Viewing pictures online can be more helpful. There are websites with thousands of pictures of even single minerals; Mindat (2020) is a good example. These can give the viewer a better idea of the range of variation seen in some minerals.

In addition to reading up on rocks, I suggest joining a local rock and mineral club. The website of the American Federation of Mineralogical Societies (AFMS 2020a) has a directory of clubs all over the US. An internet search for rock collecting clubs and geographic area can also be helpful. While there may not be a club nearby in some of the more rural areas of the country, there will likely be one, if not several, near most locales. Most clubs are full of people eager to share their knowledge with beginners. Collectively, clubs usually have centuries of experience to share.

Another advantage of joining a club is that clubs often have access to collecting sites that are otherwise inaccessible. Some clubs specialize in one aspect of the hobby, such as gem cutting or fossil collecting, though most include people with interests in all aspects. Most clubs have programs, such as speakers, that can be very educational, and some have hands-on classes, such as in gem cutting or jewelry making. Some clubs sponsor gem and mineral shows, at which people exhibit their best material and dealers offer items for sale. Shows are a lot of work for club members, but they are also lots of fun.

Reading up on rocks, and at least visiting a club, are good ways for the beginner to get an overview of the hobby, and decide if it is something that they really want to pursue. If the answer is "yes", one will already have begun accumulating the knowledge which I have mentioned above.

Chapter 3

I Specialize in Everything

Once a person decides that rock collecting is for them, that person often is faced with making a decision – whether to specialize in one or a few aspects of the hobby, or not to specialize. In my own personal case, I have chosen not to specialize. While I find some aspects more interesting than others, I basically like them all. An advantage of not specializing is that wherever I go, I can usually find or see something in which I have an interest. A big disadvantage is that I have filled my house with rocks, and have trouble finding places to display new acquisitions. More on this issue later.

Three common areas of specialization are minerology, paleontology, and lapidary (gem cutting). There are many people who pursue minerology as a hobby. Those with a background or interest in chemistry or physics are often drawn to minerals. Minerals have defined chemical compositions and physical properties. They all also have defined crystal structures, except for one, mercury, which is a liquid. Another reason that many people are drawn to minerals is that they are often simply beautiful. In my personal opinion, a well-formed mineral specimen can be as beautiful as any artwork created by humankind. There are, of course, many who would disagree, and they are entitled to their opinions.

People with an interest in biology and evolution (like myself) often develop an interest in fossils. Evolution ties together the many and varied aspects of biology, and fossils are the direct physical evidence for it. And, it can fire the imagination to look at, say, a dinosaur tooth or large shark tooth, and think of the animal to which it once belonged. Many people believe that all fossils are rare and belong in museums. While many certainly are, others are abundant. Some fossils occur by the millions, and can be had by even beginning collectors. There are certainly enough *Elrathia* trilobites from Utah, for example, to go around.

Paleontology is one field in which amateur collectors can make valuable contributions to science, and thus the generation and dissemination of knowledge. Many significant fossils have been found by hobbyists who later donated them to museums. A disadvantage of collecting fossils is that more and more restrictions on it are being put in place. The argument is that collection of fossils should be left to professionals. The counter argument is that there are not enough professionals to collect everything of value. Many amateurs search for fossils that have eroded out of their host rock, including those that wash up on beaches. If not immediately collected, these fossils would likely be destroyed by further weathering. Some fossils, dinosaur skeletons for example, may require professional expertise to properly extract and preserve. If a hobbyist finds such a fossil, it is the responsible thing to contact a professional and at least have it checked out. (More on this argument later.)

Another act of responsibility is for a collector who finds an unusual fossil, whether difficult to collect or not, is to consider donating it to a museum. I've known several collectors who have done this and been rewarded by having new species named after them. Again, this constitutes generating new knowledge. Responsible behavior by collectors can help counter the argument for more restrictions on collecting.

People with artistic talents are often drawn to gem cutting and jewelry making. I have known some amateur cutters whose work is far superior to anything found in the average jewelry store. Those with an interest in physics are sometimes drawn to faceting, as cutting designs are based in part on optics. Cuts are usually designed to return light that enters the stone to the view of the observer, which depends on such things as refractive indices. Cutters often become field collectors, with the goal of finding something that can be polished. Good cutting material can still be found in many localities, though many types are becoming difficult to find. Good faceting material, in particular, whether self-collected or purchased, is becoming hard to get. Some countries, such as Brazil, still produce lots of facet rough, but commercial cutters snap up most of it before it reaches hobbyists. Some people, as a result, cut a lot of synthetic material, and still derive much enjoyment in doing so.

One does not have to be an artist to enjoy lapidary work, however. I dabble a bit in cabochon cutting, but certainly don't consider myself to be very artistic. Just learning how to raise a polish on a bit of stone, thereby revealing its inner beauty, can be great fun. This can be as simple as using a small rock tumbler, or as elaborate as you want to make it. If one is lucky, there may be a local club that maintains a workshop and offers classes. Or, one may be able to find a club member who would be willing to teach someone individually on their own equipment.

One other bit about gem cutting. With all due apologies to cutters who may be reading this, I personally believe that some pieces should not be cut. It would be hard to beat Mother Nature's handiwork when it comes to certain well-formed crystals (e.g.,

Farrar 2005). Broken crystals often yield gemstones that are just as beautiful as do complete crystals. Again, this is just my opinion, though I know it is shared by many mineral collectors.

Another way that some rock collectors specialize is to restrict themselves to specimens that they find themselves in the field, usually called "self-collected" material. The great advantage of this is that it does not usually cost a lot of money. Some collecting sites are open to the public for no cost. The only expense might be for travel. Some privately owned collecting sites charge fees for collecting. These fees vary greatly, and can run into considerable money. One type of fee site that I personally avoid are those "gem mines" that sell buckets of dirt that have been "salted" or "enriched" with low-grade material from various sources. They may be entertaining for kids, but not serious collectors. Many excellent books listing collecting sites for various geographic areas have been published. A search of the website of any major online book seller will find them. Unfortunately, many are out of date by the time they are published.

Here is another advantage of joining a club. There are many collecting sites that are open to organized clubs but not to individuals. Most clubs carry insurance that covers damage to the property of the site owner, which helps gain access. However, club insurance does not generally cover injury to collectors. Liability concerns have led many commercial quarry owners to prohibit collecting. Those that do still allow it often impose strict rules about where and how one can collect. Again, it is the responsible thing to do to follow these rules to the letter, as this can help keep sites open for future collecting.

A big disadvantage of self-collecting is that the variety and quality of material that is available is limited. One may live in a state with limited deposits of minerals or fossils. Travel may be required to do field collecting. Some mines that are operated for gemstones or crystal specimens allow limited collecting, but are not going to allow a hobbyist to pull out a fine crystal that would pay for the operation of the mine. As mentioned, many formerly productive sites are closed. Finally, some sites are simply exhausted; everything collectible has long since been removed.

Field collectors are likely to run into a variety of laws and regulations governing rock collecting. Collecting on Federal lands, for example, comes with a variety of restrictions, including a total prohibition on collecting vertebrate fossils on Bureau of Land Management (BLM) lands (BLM 2020). There are sites on National Forest Service land that were once open but are now closed or severely restricted because irresponsible collectors went in and made a big mess. In addition to the BLM regulations on fossils, there are many other state and federal laws that affect mineral and fossil collecting. Wolberg and Reinard (1997), *Collecting the Natural World*, is an excellent summary of state and federal laws relating to the collecting of minerals and fossils, as well as plants, animals, and archaeological materials. This book is a bit out of date, but still very useful. It is unfortunately out of print, but used copies can be found online. A number of countries around the world have placed restrictions on the export of fossils; for a summary of these laws, see the website of the Association of Applied Paleontological Sciences (AAPS 2020a). In addition, there are many good practices that collectors should follow that will help ensure future access to sites, such as obtaining permission from landowners and causing no damage to properties. The American Federation of Mineralogical Societies has an excellent code of ethics summarizing best practices for field collecting (AFMS 2020b).

Sometimes, when a beginner visits a collecting site where interesting material is known to be found, it can be discouraging if they do not find something right away. Generally, if something is abundant or obvious, the majority of it will have already been removed. But, that does not mean there is nothing left. Often, it simply requires a little work and some patience. As I like to say, "persistence pays."

So, what is one to do if they want more than can be practicably self-collected? Like myself, most resort to "silver picking", i.e., buying specimens. This, of course, can run into money. Prices paid for minerals and fossils vary tremendously, from under a dollar to well over a million dollars. However, it is not necessary to spend a huge sum to build an interesting collection. If you have your mind set on something in particular, you may have little choice but to pay the price if you really want it. If, on the other hand, one has a broad interest in the field (like myself), it is usually not too hard to find something of interest at a reasonable cost. If possible, I recommend that a beginner go to a gem and mineral show. Shows are held all over the country. Chances are that there will be at least one within a reasonable distance of most places in the US sometime during the year. The free website of *Rock & Gem* magazine (Rock & Gem 2020) includes listings of shows all over the country.

Yet another reason to join a club is access to material offered for sale by members. I know several clubs that have auctions periodically, where members sell items. Members often put up for sale very nice material. Sometimes they now own even better pieces, or perhaps they are downsizing. In any case, great deals can often be had. I've seen beginners walk away with whole boxes full of specimens for just a couple of dollars. Some clubs also sponsor "rock swaps" or "tailgating", where members who may not have enough to sell to justify being show dealers can offer their material for sale or trade, again frequently for bargain prices. Club members are usually eager to help newbies, often by sharing their material, as well as expertise.

Besides the limitations on availability of self-collected material, some people resort to silver picking because they are physically unable to do field collecting. Physical restrictions, however, need not stop one from enjoying the hobby. I've known people who use wheelchairs or electric scooters who still have great fun going to shows. And, these days, there is a great variety of material for sale online. A person can build a nice collection without ever leaving their home.

Some people specialize by concentrating on one state or geographic area, whether by self-collecting, purchasing, or both. Usually, it's the person's home state, but not always. As discussed in relation to specializing in self-collected material, specializing in one's home state often comes with limitations. Some states, such as Arizona and Maine, have rich and diverse deposits of minerals, which can form the basis for a fine collection. Other states are not so rich. Specializing in rocks from a particular state would be something that a person would have to judge for themselves.

People with limited space often choose to specialize in small specimens, including "micromounts" or "thumbnails." Micromounts, technically, can include anything which requires magnification to appreciate, even big rocks with tiny crystals on them. Generally, however, micromounts are small specimens that fit in one-inch plastic boxes. Micromounting is a popular specialty; there are entire clubs devoted to it. It opens another whole world of minerals, as some species only occur as small crystals, and small crystals are often better quality than larger ones. *The Complete Book of Micromounting*, by Quentin Wight (Wight 1993) is a good reference. Thumbnails include somewhat larger specimens, basically anything that will fit inside a one-inch cube, whether it requires magnification or not. These are usually mounted in "perkies", one-inch clear plastic boxes with quarter-inch bases. Aside from the obvious saving of space, specializing in micromounts or thumbnails can also save money, as small specimens are generally less expensive than large ones. That is not to say that all small specimens are cheap; think diamonds and gold nuggets, for example. In addition, a decent microscope is a big plus, and these can also run into money.

There are other ways to specialize in addition to those I have discussed. These include individual mineral species, fluorescent minerals, radioactive minerals, etc. The list could go on and on. What I have attempted to do is to give the beginner some ideas of popular ways to specialize. One must decide for themselves whether to specialize in one of these areas, or, like me, to specialize in everything.

Chapter 4

Get Rich Quick – NOT!

In the preceding chapters, I have discussed some of the reasons that people take up rock collecting as a hobby, and I'm sure that there are other reasons out there. The point I want to make here is that these are reasons for people to pursue the *hobby*. It is not my intent to guide people who might be interested in geology as a career; that is beyond the scope of this writing. In this chapter, though, I will discuss a big reason *not* to pursue the rock collecting hobby: the desire to make money.

When visiting a rock show, it is not uncommon to see considerable amounts of money changing hands. Mineral specimens priced in the six-figure range can be seen at some shows. It is easy to get the impression that one can make a lot of money buying and selling rocks. While money can certainly be made, otherwise there would not be so many commercial dealers, it is not an easy thing to do.

Many hobbyists sell at least a few rocks from time to time. Most of us upgrade our collections as we go along. It is not uncommon for a collector to acquire a modest specimen of a particular mineral from a particular locality, only to acquire a better one from the same locality at a later time. In such cases, the collector often "flips" the old specimen, selling it to help offset the cost of the new piece. Those of us who do field collecting are sometimes lucky enough to find multiple pieces of the same thing at a particular site. Many of us pick out a few of the best pieces to keep, and sell some of the others. Sales such as these are commonly done at club auctions, directly to fellow collectors, or, these days, online. Usually, not a lot of money is involved, just hopefully enough to help pay for new pieces or travel to collecting sites. It can also help minimize the space required to store and display specimens.

Some collectors, myself included, will, when given the opportunity, buy multiple similar pieces with the intent of keeping one or two and selling the rest. Here is where one must be careful. It is easy to acquire a lot of material quickly, and to sink a lot of money into it. Recovering the investment, let alone turning a profit, requires considerable care and work. It takes experience to know what is likely to be salable, and what it is really worth. An inexperienced collector can easily acquire a lot of rocks that nobody else wants or are not worth the price paid to acquire them.

If one persists in buying parcels of material to resell, one soon has more than one can readily sell to friends and fellow club members. It soon becomes necessary establish a business, either online, as a show dealer, or both. In my case, I have become a show dealer. Here again is where care is needed. There are a lot of expenses associated with shows, including table fees, travel and lodging, supplies, helpers, etc. Of course, the best shows often have the greatest expenses, and there are usually long waiting lists to get in. (For more on becoming a show dealer, see Albrecht 2019.) Selling online generally involves fewer expenses than does selling at shows, but there are still costs (e.g., photography, packing and shipping.) I try to limit expenses by sticking to shows close to home, which are luckily numerous in my area. I have built my inventory gradually over the years, so I have not had to lay out a lot of money at once. Then there is bookkeeping. Even for a modest business like mine, there are many hours of bookkeeping required to calculate profits or losses, which I must report at income tax time. Most years, I manage to eke out a small profit. I've never tried to figure out how much this profit amounts to per hour of work I put in, but I'm sure it's very little.

Like most businesses, bigger rock businesses tend to take in more money. However, bigger businesses also have bigger expenses. For rocks, this means more inventory, a vehicle that can transport it, bigger show fees, more travel expenses, and, frequently, paying someone to help man the booth. Anyone contemplating being a big dealer should give all this careful consideration. I have, and have chosen to do business on a more modest scale.

Field collecting can also lead one into business. The thought of pulling something out of the ground for seemingly nothing and then selling it for a lot of money can be very powerful. It's easy to get "gold fever", whether one is seeking actual gold, or other minerals or fossils. It can even be tempting to acquire a piece of land and mine it for minerals or lapidary material. Here is another time to be careful. There are certainly people who have been successful mining minerals and gems. I have met some of them. But, it is a risky proposition for just about anyone. Those familiar with gem mining in Maine, for example, will know of the Perham family, who have certainly had great success mining over the years. It has not been easy even for the Perhams, though. Frank Perham has been quoted as saying "It's a quick way to go broke." His sister, Jane, was quoted as saying "Mining is worse than slot machines" (Stevens 1994.)

Some mining operations can be worked with simple tools ("pick and shovel"), which would involve limited expenses. However, mining can involve tremendous expenses in addition to the cost of the property. It may require heavy earth moving equipment, fuel, drilling rigs, and even explosives, plus permits and licenses to use it all. Not just anyone has the skills to operate the equipment either. One may have to hire skilled workers. Even with all this one may still not find anything. The bottom line is that before embarking on a mining venture, be sure you know what you're getting into, and whether you can afford to lose your investment if it all comes up dry.

There have been numerous occasions on which I have found more specimens than I need to keep, but in most cases the ones that I would be willing to part with are usually not worth much. While I have sold a few of these, I have donated to many more to clubs. I have neither the resources to do large scale mining nor the guts to risk it if I did. Well, maybe if I ever win the lottery. Otherwise, like my show business, I'm keeping it modest.

So, if I only make a piddling amount of profit selling rocks, why continue to do it? The reasons are the same as why I collect in the first place, because I enjoy it, and it helps disseminate knowledge. At shows, I usually spend more time answering questions about rocks than anything else. And, I distribute properly labeled specimens that constitute knowledge. If it ever gets to the point where it is too much work and too little fun, I will probably quit doing shows and find another outlet for my material.

My point in this chapter is not to keep people from going into the mineral business. Those of us who do silver picking greatly appreciate those who are in the business. Without them, a lot of collections would be very limited. Rather, my point is for the hobbyist to be aware of what the hobby can lead to, and to make people think about whether they want to go there. In my case, I have gone there, but I'm keeping it limited. If, on the other hand, one considers all the factors involved and still wants to go big, I say "Go for it!" Just don't expect to get rich quick.

Chapter 5

How Much is Too Much?

Like any hobby, rock collecting tends to grow over time. As we get older, we usually learn more about the phases of the hobby we are interested in, and where to find things. We often have more money to spend as we go along. And, at least when we retire, we may have more time to devote to it. Hobbies can proceed from an interest to a passion to an obsession, possibly leading to an undesirable situation. The point of this chapter is to recognize when a pleasurable pastime is getting out of hand.

As mentioned previously, humans may have an innate desire to collect things, going back to when we had to forage for food. But, in today's world, innate behaviors can sometimes cause trouble. Just as overindulging in our desire to eat can lead to obesity, overindulging in our desire to collect can lead to problems.

One often hears the term "shopaholic" applied to persons who tend to buy too much of something. That something can be rocks, whether they are purchased or self-collected. More properly termed "shopping addiction" or "oniomania," this is a recognized psychological condition (Banschick 2014). Among other symptoms, it is characterized by buying more than one needs, thrilling over purchases followed by regret, and spending beyond one's means. This can be a serious disorder that requires professional help. However, just because one finds pleasure in acquiring rocks does not necessarily mean that one is a shopaholic. It's just something to be conscious of.

Another psychological disorder to which some may suspect rock collectors being prone is hoarding behavior. Among the symptoms of hoarding are excessive accumulation of unneeded possessions, an unwillingness to get rid of stuff, and having so much stuff that it renders living space useless for its intended purpose (Mayo Clinic 2019.) (Sound familiar?) Like shopping addiction, hoarding behavior can be a serious disorder that requires professional treatment. Again, though, just having a lot of rocks does not necessarily make one a hoarder.

I'll use myself as an example. I've done a lot of field collecting over the years. When I've gotten into a productive site, I've been tempted to bring home every rock with anything interesting about it. This is like shopping, even though it may involve no money. However, the responsible thing to do in such a situation is to take only what you need for your collection, or can cut, or can reasonably expect to sell or trade. Leave some for the next collectors. It will help foster interest in the hobby if others can find material to collect. The exception is when the site is threatened with destruction, in which case everything possible should be collected. For example, I was once with a group of collectors at a site where old mine dumps were being dug up for use as road gravel. We took anything that looked interesting; otherwise, it would have soon become gravel. I found some nice things, but I also expect to wind up giving away a lot of it.

I also frequently buy specimens, and, I admit, I get a bit of a thrill out of it. But, I rarely regret buying something, except maybe if I later see a better piece for a better price. I also try to not to spend excessively. When I go to a show, for example, I usually take a certain amount of cash, and only spend that much. I occasionally write checks, but I seldom use credit cards. If I do use a credit card, I pay it off that month. I've never gone into debt over rocks. As I discussed in the previous chapter, making money off rocks is not easy. I try never to spend an excessive amount of money on rocks in hopes of making it back later when I sell some of them.

One thing that I have done that has cost quite a bit of money over the years is to travel abroad to visit famous mineral and fossil localities. I mentioned this earlier; the reader may wonder how it came about. I have written a number of articles about these trips (e.g., Farrar 2005, 2018.) However, not everyone has old issues of *Rock & Gem* lying around, so I will digress briefly. A number of years ago, I would go to Franklin, North Carolina to collect at the famous ruby and sapphire mines located nearby. (Unfortunately, most of those mines are now closed, or salted.) There I met a woman named Sara Mount, who owned a gem shop. As the mines and associated business declined, she needed to get into another line of work, so she decided to become a travel agent. Her interest in gemstones led her to start doing trips to famous mineral localities. She initially developed connections in Brazil, and, later, in Morocco and Thailand. I have gone on a number of her trips to Brazil and Morocco, and one to Thailand.

Of course, international travel does not come cheap. Plane tickets can run into money. Lodging in some countries can sometimes be had cheaply, but Sara likes to utilize reasonably nice accommodations. We do not "rough it" on these trips; we always have a roof over our heads and indoor plumbing at night. I may enjoy getting hot and dirty visiting a famous gem mine, but I like to have a good shower and comfortable bed afterward. On these trips, I frequently buy specimens for my collection. I can often get them cheaper than I could in the US. Also, I often see rare or unusual items that would be hard to find at shows. However, I do not save enough on specimens to pay for the trips. I could probably take the same amount of money to a big rock show and come home with more stuff. The reason that I go on such trips, rather, is to learn about localities from which the material comes, as well as to experience other cultures and see interesting historic sites and natural areas along the way. I gain quite a bit of knowledge on these trips, much of which I have attempted to disseminate through published articles (e.g., Farrar 2005), and talks to various clubs. Again, while travel entails spending money, I have never spent money that I did not have.

All of this collecting, shopping, and traveling has led me to accumulate a lot of rocks. I do, however, strive to avoid becoming a hoarder. One of the characteristics of hoarders is an unwillingness to get rid of stuff. I have willingly gotten rid of a lot of rocks over the years. Some of these rocks I have sold, but many I have given away or donated to clubs. I have probably given away thousands of rocks over the years. I frequently go through my rocks to decide which ones I can get rid of. The one thing that I am reluctant to do is simply throw them in the trash. That would entail loss of the knowledge that I have mentioned frequently. No, I much prefer that they be passed on to someone who will be interested in them, and will enjoy having them.

Another characteristic of a hoarder is that their stuff takes up so much of their living space that it renders that space useless for its intended purpose. I admit that I have a lot of rocks around, but I try to keep them organized and in boxes, drawers, or display cabinets, or at least out of the way. I do not let them block egress; that could be dangerous in the event of a fire.

Sometimes, spending time and/or money collecting rocks can lead to conflicts with family members. I once saw a bumper sticker on the car of a model railroader that said something to the effect of "My wife says that if I buy one more train, she is going to leave me. I'm sure going to miss her." Just substitute "rock" for "train," and you have some people I've met. Relationships with family members are different for every person, but are something that one should always consider.

While it is possible for a rock collector to become a shopaholic or a hoarder, I suspect that serious problems are uncommon. These are just cautionary notes, possibilities to be aware of. I still would encourage people to pursue rock collecting as a hobby. But, just like everything else we do, be considerate of how it affects your family, and, of course, live within your means.

Chapter 6

Ethical Dilemmas

I like to think that most rock collectors are decent folks who care about the welfare of others. And, most decent folks do not want to do anything that could contribute to harm to other people or to the environment. For example, I always try to recycle whatever I can, and I would not knowingly buy stolen items. Some may wonder about potentially harmful effects associated with rock collecting. This is a legitimate concern, and something that should be considered in just about any activity.

To an environmentalist, the very mention of any type of mining brings to mind serious environmental issues. There have certainly been far too many cases in which mining has caused great damage to the environment. Mining has been responsible for large scale habitat destruction, toxic runoff, siltation, and pollution. Rock collectors may legitimately be concerned that their activities might contribute to such problems. However, in most cases, the environmental impact of hobbyists is minimal.

Most of us who do field collecting are incapable of doing much damage to the environment. In most cases, working with picks and shovels is not going to cause significant environmental degradation. There are, of course, exceptions, including environmentally sensitive areas that are home to rare plants or animals, where collecting is unwise, and likely illegal. One should investigate to see if such sites are in a particular area before doing any collecting. In most cases, however, holes that can be dug by hand are not great environmental threats. If one decides to move from hand tools to power equipment, though, potential environmental effects may need to be considered. For additional guidelines that collectors should follow to minimize their impact, see the aforementioned AFMS code of ethics (AFMS 2020b)

There are many commercial mines, both in this country and around the world, that are operated primarily for gemstones and/or mineral specimens. At many of these sites, power equipment, and possibly explosives, are used. Earth moving equipment, of course, has the potential to cause environmental harm. Compared with mines operated to produce industrial minerals, such as metal ores, coal, or building stone, however, most gemstone or specimen mining operations are quite small. I have seen famous gemstone and specimen mines that only occupy a few acres (e.g., Farrar 2005, 2018). The environmental impact of such small operations may not be zero, but compared to industrial mining it is usually minimal.

Some of the mineral specimens that we see for sale at shows come from industrial-scale mines that do have the potential to cause significant environmental problems. Some of these may have been collected by mine workers apart from their regular duties. More and more mining companies, though, forbid their workers from collecting mineral specimens. I hate to think how many fine crystals are destroyed in the pursuit of industrial minerals. Even if mineral specimens are recovered from industrial mineral mines, they would usually constitute a minor by product.

Not purchasing these minerals at shows would have little if any effect on the operation of the mine itself, and would do little or nothing to prevent associated environmental degradation.

Perhaps a more serious concern than how specimen mining affects the environment is how it affects people. We have all heard the terms "blood diamonds" or "conflict diamonds", referring to diamonds mined by forced labor to finance armed insurgencies. I certainly would not buy a diamond if I thought that it came from such a source. On the other hand, I have bought specimens directly from miners in this country without a second thought. There are, however, many sources of specimens that fall somewhere in between.

One often hears the term "artisanal mining" used in reference to small mining operations. "Artisanal", in everyday parlance, has a positive connotation, and usually refers to high-quality products made in small quantities (e.g., "artisanal cheeses".) Artisanal mining, however, usually has a negative connotation, and refers to any small, often illegal, mining operation, including blood diamond production. Most artisanal mining focuses on industrial minerals, such as cobalt, tin, gold, tantalum, and diamonds. In some countries, thousands of people labor under miserable conditions to recover these commodities. Middlemen often get most of the profits, with little going to the miners themselves. People are not always physically forced to do this kind of work as may be the case with blood diamonds, but economic conditions in some countries may force them into it. For more on artisanal mining, see Voynick (2019).

Many mineral specimens, gemstones, and fossils are found in operations that are also considered artisanal mining, but do not necessarily involve the miserable working conditions often seen in industrial mineral production. I have personally seen many such operations in my travels to Brazil and Morocco. Most operations that I have seen involve one, or a few, people working at a given site. In Brazil, they have a word for these people: *garimpeiros*. Most of these miners are poor by US standards, though I have met some who have found pockets of crystals worth tens of thousands of dollars. In their eye, the chance to make such a find may beat the limited economic opportunities that may exist in the area. Most artisanal miners operate with simple tools, without the safety measures required in mines in the US, so there is physical risk involved. I know that in Morocco, for example, people have been killed in cave ins while digging for fossils.

One may be concerned that buying a mineral specimen from an artisanal mine supports the exploitation of the workers. This may be the case with blood diamonds and some other items, but in the artisanal gem and specimen mines that I have seen, workers are not forced to work there. They usually make the choice themselves. Buying specimens from these operations provides economic support that they might not otherwise have, especially if it can be done directly, rather than through middlemen. I once knew a woman who had spent time with the Peace Corps in Africa. There, she had bought a diamond from a local man, but later felt guilty about exploiting him. I told her that if she had not bought it, he would probably have sold it to a dealer, who would probably have given him less for it than she did. As it was, they both came out ahead; she got a great souvenir, and he got more money. Most collectors do not have the opportunity to buy stones directly from garimpeiros, but the money they spend at shows may eventually find its way back to them.

I have frequently mentioned the knowledge that specimens constitute, and I will do so more later. While many may be conscious of potential effects of collecting on the environment or people, some may also be aware of potential effects on scientific knowledge. This most often comes up in connection with fossils.

As I mentioned before (Chapter 3), there is a long running debate over whether private collection of fossils helps or hinders the science of paleontology. Some argue that collection of fossils should be left to professionals. Others argue that there are not enough professionals to do all the work. Many would argue that it is better for a rare fossil to be in a private collection than for it to remain buried forever or destroyed by erosion or mining for industrial minerals.

Most fossils offered for sale at shows or online are well-known to science, and diversion of some specimens to private collections would be of little scientific consequence. This is particularly true of specimens that are within the budget of the average collector. Some of the more common Moroccan trilobites, such as *Colpocoryphe* (often called *Calymene*), are a good example. Thousands of these are found every year, and they can often be had very cheaply. Even some pricey specimens are quite well documented. For example, large *megalodon* shark teeth in good condition may sell for high prices, but they are well represented in museums.

The picture gets murkier when it comes to rare or unique specimens sometimes found by private collectors. Some wealthy collectors will pay large sums of money to acquire spectacular fossils that museums would like to have. Some of these are indeed lost to science, but others do eventually find their way into scientific study. A rich collector's heirs, for example, may prefer to donate pieces and take a tax write-off than to keep the fossils.

Though the debate is not likely to be resolved any time soon, there seems recently to have developed increasing cooperation between the two sides. In a recent article in *National Geographic*, Conniff (2019) discusses the fact that, given recent budget cuts to museums, professionals may need amateurs. Hopefully this spirit of cooperation, grudging though it might be, will continue.

As I have said, if a collector has something that seems unusual, the responsible thing to do is to at least show it to a professional. In the past, many scientific journals have been reluctant to publish papers written about specimens that are not housed in museums, but this is starting to change. Who knows; it may turn out to be a new species and would make a valuable contribution to science, and it could help ease the conflict between amateurs and professionals.

For more on the ethics of fossil collecting, see the website of The Association of Applied Paleontological Sciences (AAPS 2020b). They publish a code of ethics regarding the collection of, and trade in, fossils, which covers many of the same points I have attempted to make here. They also publish a summary of international laws regarding the fossil trade (AAPS 2020a). Additional information on US laws is summarized by Wolberg and Reinard (1997).

A good example of a material with multiple ethical dilemmas associated with it is amber from Myanmar (formerly known as Burma). Not only is Myanmar amber a beautiful gemstone, it frequently preserves fossils, including a myriad of insects, as well as small vertebrates, such as small birds, snakes, and even small dinosaurs. This amber was the subject of a recent paper in the journal *Science*. I'm sure that most readers do not subscribe to *Science*, but a version that anyone can view was published online by Sokal (2019). To begin with, most Myanmar amber has been mined under dangerous working conditions. As with blood diamonds, profits have been used to finance an armed insurgency in northern Myanmar. Much of the material is smuggled out through China, where it is sold to anyone with money. Some is purchased by scientists, but much goes to private collectors. The government of Myanmar recently took control of the mining area, and the larger mines have closed, so the supply of amber may be ending, but one can still find it for sale. Clearly there are ethical issues here. Most people do not want to finance an armed insurgency, and many of the fossils that have been found belong in museums. However, it could be argued that it is better that the amber that contains fossils go to responsible collectors than for it to made into jewelry. (There is lots of amber that does not contain fossils that can be used for jewelry.) Not buying fossils is not necessarily going to leave more for scientists, who often have limited budgets. One would hope that valuable amber fossils purchased by collectors would eventually wind up in museums, as with dinosaur fossils discussed above, but this may not always be the case.

The purpose of this chapter is to highlight some of the ethical questions that a rock collector may face at times, and to present some associated arguments. It is not to answer these questions; that is something each person must do for themselves. However, as the reader may surmise, I, personally, like most collectors, come down on the side of continuing to collect, but with the awareness that there may be some situations where collecting is not ethical.

Chapter 7

Label, Label, LABEL!

I have written about knowledge many times so far, and that is the primary focus of this chapter. I have said that properly labeled mineral and fossil specimens constitute knowledge, i.e., the knowledge that that particular mineral or fossil occurs at a particular place. The key words here are "properly labeled." The main thing that constitutes proper labeling is accurate locality information. A specimen can always be identified later, but missing locality information is often gone forever. A specimen of, say, calcite, with no locality information does not constitute knowledge that calcite occurs at a particular place. That calcite specimen may still be useful for educational purposes, such as demonstrating double refraction, or as a decorative piece to sit on a shelf somewhere. But, it does not constitute knowledge.

Some people may think they don't need to label everything, that they will remember where it came from. For a beginner with a few specimens, this may be easy, but the more we acquire, the harder it gets. And, as we get older, memories tend to fade, and of course are lost forever when we die. So, like most collectors, I recommend labeling everything.

What information on locality is need on a label? Generally, the more detailed the better. The name of the mine, if it is from a named mine, is very important. Neighboring mines can be very different. For example, at Mt. Apatite, near Auburn, Maine, there are several mines within a few hundred yards of each other, each with distinctive mineral assemblages. One of these mines is the Pulsifer Quarry, perhaps the most famous locality for purple fluorapatite in the world. A specimen labeled "Pulsifer Quarry" thus carries more information than does one labeled "Mt. Apatite" or "Auburn." Pieces from classic localities like the Pulsifer Quarry are also often worth more than similar pieces from other localities (if similar pieces exist, that is.) Some specimens come with more detail about the mine, such as level (e.g., "500-foot level"). Some miners give names to each crystal pocket. For example, I have a specimen of rhodochrosite from the "Tetrahedrite Stope" of the Sweethome Mine, Alma, Colorado. If available, such details should be included on the label.

Aside from the name of the mine, if known, geographic area is of most importance. If a specimen does not come from a named mine, e.g., was found while hiking in an undeveloped area, geographic area is especially important. In general, this includes the nearest incorporated town and the state, or province. This is usually standard for specimens from the US, and often Canada. In most cases, if the town and state are known, it is not necessary to include the county, as this is easy to find on a map. For specimens from outside the US or Canada, it is imperative to include the country, if known. A label with "Tsumeb, Namibia" on it, for example, is much more informative than one with just "Africa."

Getting the proper name for a country for a label can sometimes be confusing, especially for older specimens, because the names of some countries have changed over the years. For example, the Democratic Republic of the Congo was known for a number of years as Zaire, and Myanmar was long known as Burma. However, with today's online resources it is usually easy to find the current name of a country. I admit that some of my labels do not have the current name of the country, but with the information that they do have, I could easily find it. If pieces are to go on public display, though, the most current name of the locality should be used.

A term that we sometimes hear with specimens is "provenance." Generally, provenance includes information not only about where a specimen was first collected, but also about who has owned it. Provenance is often indicated by having old labels with the name of the collector or dealer on them. For example, I have a small aragonite specimen with four labels, going back to Adam August Kranz, in Germany in the 1800's. While knowing who has owned a specimen may not contribute much to scientific knowledge, it does contribute to historic knowledge. It also increases the value of a specimen. This is particularly true of more expensive pieces sometimes seen at shows. However, it can increase the value of any specimen, such as my aragonite. If you acquire a specimen with one or more old labels, these labels should be kept with the specimen, especially if they include the name of the previous owner. However, a label scrawled on a scrap of paper with no information on owners does not add much if kept. In these cases, I usually make a new label and discard the old one. I would suggest adding your own label to all specimens; who knows – maybe you too will become a famous collector someday. For more on the importance of provenance of mineral specimens, see the editorial by David Trinchillo published in *Mineralogical Record* (Trinchillo 2015.)

In most cases, if locality information is lost, it is gone for good. In some cases, though, it can be reconstructed. There are some minerals that are only known from one locality. Well-formed benitoite specimens, for example, have only been found in San Benito Co., California. Specimens of some minerals from some localities have distinctive appearances. Yellow brucite specimens recently found in Pakistan are one example. However, in most cases, recognizing the source of a piece requires a great deal of experience, and is not always reliable even with experience. Therefore, every effort should be made to preserve original locality information.

Like locality names, the name of the specimen may change, which can lead to some frustration. This seems to be especially true for fossils. Take the *megalodon* shark for example. When I first learned about it, it was known as *Carcharodon megalodon*, in the same genus as the modern great white shark, *C. carcharias*. Later, it and its relatives were moved to the genus *Carcharocles*. And, most recently, I have seen it referred to as *Otodus megalodon,* placing it in the same genus as its possible ancestor, *O. obliquus* (reviewed by Kent 2018). As scientists find new specimens and learn more about fossil species, such name changes are bound to continue.

Likewise, mineral names can change. The mineral commonly known as apatite is a good example. Many years ago, it was divided into three mineral species, fluorapatite, hydroxylapatite, and chlorapatite. These names were later changed to apatite-(CaF), apatite-(CaOH), and apatite-(CaCl), respectively. Then, much to everyone's consternation, they were changed back to fluorapatite, hydroxylapatite, and chlorapatite. Figuring out the proper name of a mineral can sometimes be challenging, especially for specimens with very old labels. The "Glossary of Obsolete Mineral Names", freely available online (Bayliss 2012,) is an excellent resource that can aid in figuring out correct mineral names. A standard reference is *Fleischer's Glossary of Mineral Species* (Back 2018). Updated every few years, it lists the correct names of all approved minerals at the time of publication. Though it may take some effort, if you have an old name for a fossil or mineral, it is generally possible to find the current name. As with locality, it is especially important to have the correct name of anything put on public display.

Most collectors use paper labels kept with their specimens, which is fine as long as they stay together. Some collectors, and many museums, attach a small label with a number on it directly to the specimen. They then put that number on the label, which aids in keeping them together. Some collectors keep additional information elsewhere, with the number to match it up. In the past, many people used paper index cards for that information, but now digital spreadsheets are often used. This has the advantage of allowing one to record more information than can easily be put on the label, such as the date acquired, cost, dealer name, etc. As I have repeatedly written, preserving information is important. The problem with these systems comes if and when the collection is broken up (more on this in the next chapter.) Even some museums occasionally deaccession some pieces. The specimens often then pass to other collectors without the catalog. Unless the catalog number is on the label, it is often of no use to later owners. For example, I have a smoky quartz crystal that has had at least four different catalog numbers attached to it, but I have no corresponding information on any of them. At least I have a label with locality information, probably prepared by the fifth or later owner. Therefor, while having a catalog may be a good idea, it is also important to have an accurate label with each specimen.

I keep most of my specimens in small paper boxes with the labels attached to the boxes, or in clear plastic boxes with labels attached inside. However, I also have some large pieces that do not fit conveniently in display boxes. Some of these are on shelves, or the floor, or anyplace I can find room for them. In this situation, it is easy for the labels to become separated from the specimens. I therefor attach small labels directly to these specimens. These labels I write in indelible, fade-resistant ink on plastic and glue them to the pieces in inconspicuous spots. I also make standard paper labels for people to read, so that it is not necessary to turn the rock over to see the information.

While I sometimes hand write labels, as I do for large specimens, I generally make my labels on the computer and print them out. My penmanship has always been poor; some of my grade school teachers used to describe my cursive as "chicken scratch." And it only seems to have gotten worse as I've gotten older. Thank goodness for computers and laser printers. Some people have beautiful handwriting, though, and can easily hand write legible labels. However you make your labels, just be sure that they are legible; otherwise, they are of little use.

In addition to individually labelled specimens, many collectors, myself included, have boxes, flats, buckets, etc. of bulk material stored away. In my case, much of this is self-collected material. It is equally important to label such material as it is to label individual pieces. Most such boxes that I have contain material from a single locality. For these, two labels with at least locality information are sufficient until individual pieces are prepared. One label goes on the outside of the container. But, because these labels may fade and/or fall off over time, it is useful to have a second label inside the box.

Locality information is often not kept with gemstones, but it can be important. Gemstones from certain localities are often more desirable to collectors, and thus command higher prices. An emerald from North Carolina, for example, will fetch a higher price than one of equal quality from Brazil. It is difficult to keep a label with a stone set in jewelry and being worn, but the information should be recorded somewhere.

While I consider some information, particularly locality, to be essential on labels, other information is optional. Some collectors like to put the chemical composition on the label, while some like to put the crystal system. This is fine, as long as it doesn't get too messy. A specimen with more than one mineral on it might have a lot of chemical and crystal information to go with it. Such information, however, is easy to find if the identification is accurate. *Fleischer's Glossary of Mineral Species* gives composition and crystal system for all known minerals, and similar information is also easy to find online.

Another bit of information that some collectors include on their labels is the date collected. This information is not usually included on specimens bought from dealers. For self-collected material, though, it is usually known and can be put on labels. While less critical than locality, it does add to the scientific and historical knowledge represented by the specimen.

Having proper labels for specimens is thus of paramount importance for the knowledge that minerals and fossils constitute. This segues into the topic of my final chapter, preserving that knowledge for the future.

Chapter 8

The Future: Don't Let Your Collection Die with You

The running theme of this book has been the generation, preservation, and dissemination of knowledge. The collecting, labeling, buying, selling, and donating that many of us do during our lifetimes contribute to this goal. It is an unfortunate fact, however, that our lifetimes are finite, and of unknowable duration. I have heard it said that we are only temporary custodians of specimens. I tend to agree. That philosophy brings up the importance for the responsible collector to ensure that their specimens and the knowledge they represent are passed on to the next custodians when that collector passes.

In all too many cases, collectors die without making arrangements for the disposition of their collections. The collector's family may have no knowledge of, or interest in, minerals or fossils. They may not realize the scientific or monetary value of them. I shudder to think how many rock collections have been thrown away when their owners died. It is thus the responsible thing to do for collectors to prepare instructions for their survivors telling what should be done with their collections when they are gone. I have done so myself.

Ideally, there might be someone in a collector's family that would like to take over the collection and continue to build it. However, such is often not the case. Nevertheless, there are many options for the responsible disposition of a collection. It might be donated to a museum, school, or club, sold to a dealer, or auctioned off. The important thing is for the collector's survivors to know the collector's wishes and how to carry them out. Contact information for potential recipients is especially important. This could be written into the collector's will, or in other written instructions given to one's heirs.

Many of us would like our specimens to go to a museum. However, museums are often very selective about what they will accept. Most museums have limited space and staff, so they can't take everything. Museums may want high end specimens that often only well-heeled collectors can afford, such as showy fossils mentioned above. They may also want unusual pieces, such as unique fossils which amateur collectors sometimes find. Something that a collector should be aware of, however, when contemplating donating specimens to a museum is that the museum may want documentation that the specimens were collected legally. They may not accept, for example, a fossil from a country that prohibits the export of fossils. This is something that collectors should investigate before putting museum donations in their wills.

The same thing goes for schools. Although schools might be interested in lower grade pieces than museums might, they also usually have limited space and personnel. There may be tax advantages in donating material to schools or museums; a tax professional should be consulted for advice on this matter.

Another option for donations is to donate material to local gem and mineral clubs. As mentioned in Chapter 2, there is likely to be at least one nearby club in most parts of the country (AFMS

2020a). Clubs are generally going to be less picky than museums or schools; they will often take anything. The club can then sell the material, either to members or the public, to help defray the costs involved in maintaining the club. They may also use it for educational purposes at shows and other events. Many club shows have kids' tables with giveaways for youngsters; kids are often thrilled to get even small, broken bits. Most pieces donated to clubs find their way into the hands of other collectors who appreciate them and will become the next custodians. This is a good way to preserve and disseminate the knowledge that specimens constitute.

In some cases, a collection may have a great deal of monetary value, and the collector may wish for his or her family to benefit financially from it. If the collector has to go to a nursing home or other facility, they may need the money themselves. This entails selling the collection. In many cases, it can be sold to a dealer. Unless the family is familiar with the business, the collector should leave instructions giving contact information for possible buyers. Another option is to hold an estate auction, or, in the case of a person moving to assisted living, a "living estate" auction. For high end material, a commercial auction house may be appropriate. For most of our collections, though, arranging an auction through a club may be more appropriate. I live in an area with many nearby clubs, several of which have held estate auctions at one time or another. One advantage of going through a club is that there will usually be members who know the material, and how to get the most for it. The club usually keeps a small percentage of the revenue, with the rest going to the family. The specimens, in turn, usually go to collectors who are happy to have them and will take good care of them.

The family should be aware that whether the collection is sold to a dealer or auctioned off, they may not get back all of the money that the collector may have spent building the collection. Certainly, some pieces may have increased in value and may bring more than the original price, but others may not. A dealer has to make a profit when he sells the pieces, and auctions may or may not bring high prices, depending on the buyers that show up.

The point of this chapter is that collectors should plan for what happens to their collections after they are gone, ensuring that their specimens will go to people who will appreciate them and preserve the knowledge that they constitute. Don't think that just because you may be young you can put this off. We never know when we might get hit by a bus or something. Plan not only for your own future, but also that of your collection.

Afterword

I have presented in this book some of my thoughts and philosophies relating to something that has been a big part of my life, the hobby of collecting rocks. I know that there are many others out there, collectors of rocks, or other things, who have had similar thoughts. Of course, everyone is different, and will have their own thoughts and philosophies. I just hope that some of my thoughts will be helpful when you get the collecting bug.

Bibliography

AAPS 2020a. International Laws Pertaining to the Collection, Import, Export and Sale of Fossil Material. Association of Applied Paleontological Sciences. https://www.aaps-journal.org/International-Fossil-Laws.html. Accessed 19 February 2020.

AAPS 2020b. AAPS Commercial Paleontology Code of Ethics. Association of Applied Paleontological Sciences. https://www.aaps.net/ethics.htm. Accessed 19 February 2020.

AFMS 2020a. Club Sites. The American Federation of Mineralogical Societies. http://www.amfed.org/club.htm. Accessed 20 February 2020.

AFMS 2020b. Code of Ethics. The American Federation of Mineralogical Societies. http://www.amfed.org/ethics.htm. Accessed 20 February 2020.

Albrecht, Donna. 2019. Getting Into the Business. Rock & Gem. December 2019. Pp. 56 – 61.

Back, Malcolm E. 2018. Fleischer's Glossary of Mineral Species

2018. Mineralogical Record. Tucson, AZ.

Banschick, M. 2014. The Shopaholic: When Shopping Becomes an Illness. Psychology Today. https://www.psychologytoday.com/us/blog/the-intelligent-divorce/201407/the-shopaholic . Accessed 9 March 2020.

Bayliss, Peter. 2012. Glossary of Obsolete Mineral Names. https://mineralogicalrecord.com/journal.asp . Accessed 9 March 2020.

BLM. 2020. Rockhounding. Bureau of Land Management. https://www.blm.gov/basic/rockhounding Accessed 9 March 2020.

Catchpole, Heather. 2010. My Brain Made Me Buy It. ABC Science. https://www.abc.net.au/science/articles/2010/12/06/3080675.htm Accessed 9 March 2020.

Conniff, Richard. 2019. The Dinosaur in the Room. National Geographic. October, pp. 124 - 141

Farrar, Bob. 2005. Caldrons of Garnets. Rock & Gem. August, pp. 88 – 91.

Farrar, Bob. 2018. Moroccan Aragonite: Specimens by the Millions. Rock & Gem. August, pp. 58 - 63.

Kent, Bretton W. 2018. The Cartilaginous Fishes (Chimaeras, Sharks, and Rays) of Calvert Cliffs, Maryland, USA. pp. 45 – 160 In Godfrey, Stephen J. (ed.) The Geology and Vertebrate Paleontology of Calvert Cliffs, Maryland, USA. Smithsonian Contributions to Paleontology, vol. 100

Mayo Clinic. 2019. Hoarding Disorder. https://www.mayoclinic.org/diseases-conditions/hoarding-disorder/symptoms-causes/syc-20356056 . Accessed 9 March 2020.

Mindat 2020. https://www.mindat.org/. Accessed 9 March 2020.

Pough, Frederick H. 1997. A Field Guide to the Rocks and Minerals. 5th edition. Houghton Miflin, Harcourt Publishing Co., Boston, MA 480 pp.

Rock & Gem. 2020. https://www.rockngem.com/ Accessed 9 March 2020.

Sokal, Joshua. 2019. Troubled Treasure. https://www.sciencemag.org/news/2019/05/fossils-burmese-amber-offer-exquisite-view-dinosaur-times-and-ethical-minefield. Accessed 9 March 2020.

Stevens, C. J. 1994. Maine Mining Adventures. John Wade Publisher. Philips, ME. 209 pp.

Trinchillo, David. 2015. Provenance: Its importance in Mineral Collecting. Mineralogical Record. Vol. 46, No. 2, January – February, pp. 212 – 213.

Wolberg, Donald, and Patsy Reinard. 1997. Collecting the Natural World: Legal Requirements & Personal Liability for Collecting Plants, Animals, Rocks, Minerals, & Fossils. Geoscience Press, Inc., Tucson, AZ. 330 pp.

Voynick, Steve. 2019. Artisanal Mining. Rock & Gem. May, pp. 26 – 33.

Wight, Quentin. 1993. The Complete Book of Micromounting. Mineralogical   Record,   Tucson,   AZ.   283   pp.

ABOUT THE AUTHOR

A native of Tennessee, Bob Farrar is a writer and a longtime collector of minerals, fossils, and gemstones, and has traveled extensively in pursuit of the hobby. He is active in several Maryland clubs, and is a dealer at area shows. He holds a Ph. D. in entomology and is retired from the US Dept. of Agriculture in Beltsville, MD.

www.ingramcontent.com/pod-product-compliance
Lightning Source LLC
Chambersburg PA
CBHW070224180726
47999CB00017B/2175